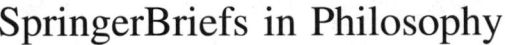SpringerBriefs in Philosophy

For further volumes:
http://www.springer.com/series/10082

Peter Kosso

A Summary of Scientific Method

 Springer

Peter Kosso
Department of Philosophy
Northern Arizona University
Flagstaff, Arizona
USA
e-mail: Peter.Kosso@nau.edu

ISSN 2211-4548   e-ISSN 2211-4556

ISBN 978-94-007-1613-1   e-ISBN 978-94-007-1614-8

DOI 10.1007/978-94-007-1614-8

Springer Dordrecht Heidelberg London New york

Cover design: eStudio Calamar, Berlin/Figueres

Printed on acid-free paper

Springer is part of Springer Science+Business Media (www.springer.com)

Preface

Scientific literacy is an essential component of any education in our scientific society. And just as linguistic literacy requires more than learning a lot of words, scientific literacy requires more than learning a lot of facts. Literacy includes knowing the grammar, and scientific literacy requires understanding the structure of scientific method. This is our goal, understanding the scientific method.

It may seem pretentious, or simple-minded, to presume that there is a single method for all the sciences. Some will suggest that using the definite article, as in THE scientific method, is inappropriate. And yet, important debates about the propriety of teaching intelligent design or creationism in public schools, or the credibility of evidence for global warming, often turn on whether or not these things are scientific. So there must be enough substance in the general category of being scientific to support these arguments. It is that substance that we will distil here.

Is the theory of intelligent design scientific or not? Well, we can't even begin to answer this question, at least not in a reasonable and profitable way, without a clear understanding of what it is to be scientific. There must be something shared by all the sciences that makes them scientific, and it would be this something that is missing from the unscientific or the pseudoscientific. That something is not what they study. Geology, biology, and physics study pretty different things, whereas biology and intelligent design study pretty much the same thing. What is common to the sciences is the basic structure of how they study, and the standards they use to judge acceptable results. This is the scientific method.

So, we will presume that there is a shared method for all the sciences, and that it is the method that makes them scientific. Furthermore, we will presume that the criteria for being scientific are objective criteria. That is, it is not a matter of personal judgment as to what the scientific method is or what qualifies as scientific. There are impersonal, objective standards for what it is to be scientific, and there is an objectively accurate description of the scientific method. This, too, must be assumed by the debates about difficult social issues like intelligent design or global warming, or else there would be no point in trying to establish one thing to be, or not to be, scientific.

Our goal is to clarify these standards and give an accurate description of scientific method. In other words, we will answer three related questions: What is science? What is the scientific method? What is scientific about science?

Acknowledgments

I have presented some of these ideas about scientific method in talks and seminars with colleagues in the sciences. I thank them all for their patience and helpful comments. Ron Markle, in particular, provided both opportunities and encouragement. Ralph Baierlein commented with helpful insight and precision on an earlier draft of the manuscript.

Introduction

This description of the scientific method fits into a discipline called philosophy of science. The goal is a clear answer to the question, What is science? But it would be wise to start with a clarification on the approach and perspective to be taken. That is, it will help to first answer, What is philosophy of science?

First, what is philosophy? It is the forum in which fundamental concepts and claims that are taken for granted in other disciplines and life on the street are questioned and clarified. Surely some things must be assumed and used without question in our academic and daily lives, but just as surely there should be a place where those things are openly discussed and questioned. That's philosophy. It may be about relations between people, where we rely on concepts like justice, morality, rights, and even what it is to be a person. Or it could be about relations between people and nature, as in the concepts of knowledge, truth, observation, and evidence. Or it could involve relations within nature itself, such as cause and effect, space and time, and laws of nature. These are all fundamental to our intellectual lives, and we operate with a presumption that these ideas are clear and that there are some basic truths about them. The business of philosophy is to make sure we understand these important concepts and that the things we assume about them are true. In some cases we find that we don't really understand or that there is no basis for our assumptions. This is worth noting, and where possible, worth fixing.

Philosophy of science does this with fundamental concepts of science. The taken-for-granted concept begins right away with the word "science". What is it? What do all those things we call science have in common? What makes science scientific? There might be an easy answer to this, but it's worth making sure and making it explicit.

Usually, the answer to this initial question invokes a reference to scientific method, and that requires some manner of empirical testing. Again, this may be easy and true, but again it's worth making sure we are clear on what the method is. Somebody needs to ask about the details of this testing, and make sure it leads to good reason to believe the results. Along the way, more basic ideas emerge, things like evidence, experiment, prediction, hypothesis, theory, law, and so on. If science is important and worthy of respect, then someone needs to see that it's clear what these things are, how they fit and function together, and how (or if) it all adds up to a way, perhaps the best way or the only way, of understanding nature. That's the goal of philosophy of science.

Science and philosophy have some important things in common. Different sciences, like chemistry, geology, and biology, share a basic method. They differ in what sort of thing, what aspect of nature, they study. And then, details in applying the method will differ according to the demands of the topic. The sciences have distinctive names, but they could be known as the science of this and the science of that. Similarly, different aspects of philosophy have a basic method in common, a shared form of analysis. They differ in what they study. There is philosophy of science, philosophy of ethics (simply called ethics), philosophy of art, of mind, of this and of that. In this way, philosophy is not all that different than science.

One significant difference between science and philosophy is that science is often rewarding while philosophy often frustrates. This is because science generally starts with what we don't understand and explains it in terms of what we do understand. Why do astronauts appear weightless? Well, you know about centripetal forces and what holds the spaceship and its contents in orbit, so ... Philosophy, by contrast, often starts with what we do understand, or at least think we understand, and reveals that it's not as simple or as clear as we thought. Everyone knows what knowledge is. That's almost a self-fulfilling statement. And yet one of the central concerns of philosophy is epistemology, the study of knowledge itself.

Sometimes, though, science starts with what's obvious and shows it to be complicated. Ask a beginning physics student what they are studying and they are likely to tell you they've spent a long time studying motion. As it should be, according to Aristotle, the Fourth Century B.C. Greek philosopher and scientist who claimed in the *Physics* that "Without understanding motion we could not understand nature" (Leonardo da Vinci went even further, claiming that an understanding of motion is not only necessary to understanding nature, as is implied by Aristotle, but also sufficient. He said that, "To understand motion is to understand nature."). But who doesn't understand motion? A close look at the concept and the evidence, though, shows details and fundamental properties that are not at all obvious. It is exactly that sort of close look at a phenomenon that may initially seem clear, in this case the phenomenon of science itself, that we will use to understand the scientific method.

Our focus will be on just one of the many topics addressed by philosophy of science. It is worth explicitly mentioning one of the topics to be skipped, scientific realism. The question of realism asks whether the best scientific results show theories to be true or, no less respectably, simply the most practical guides for dealing with nature. A clear account of scientific method leads right up to this question and prepares it for a clear debate. The difference between the methodological work to be done here and the challenging issue of realism is a difference between describing and evaluating science. Our job is to build the model of how science works. Hopefully, that model can then serve as a starting point for discussions about realism.

Contents

Chapter 1
Science and Common Sense

Abstract Scientific method is not very different than what everyone does on a daily basis in coming to know about the world. Respect for evidence and reason are basic common sense and basic scientific method. So, disregarding scientific standards and results in selected aspects of life amounts to disregarding common sense. This is not to say that scientific method is perfect. It is important to note the limitations and potential misuse of the method. Unrealistic expectations will turn any imperfection into a reason to discount the whole thing.

Keywords Evidence · Reason · Uncertainty

What makes science scientific? What distinguishes science from other things people do or study? It's not about the sorts of things science studies. It's about how it studies things. The term "scientific" usually describes a process or the standards required of a process, not any kind of object or category of things. If you free-associate on the word "scientific", the most likely follow-up is "method". It sometimes even gets a definite article, as in THE scientific method. And the basic ingredients of the scientific method are familiar to most of us. Some mix of observation, evidence, testing, and logic is required, or whatever you have done, it's not scientific.

Scientific method is not much different from our day-to-day ways of learning about the world. Without really thinking about the steps or the standards, common sense invokes the same process of evidence and reasoning as scientists more explicitly follow. Your cereal is in the bowl and something about the milk doesn't seem right. You suspect the milk is spoiled. This is a hypothesis. So you check the date on the carton, and you smell the contents, and sure enough, the milk is old and sour. This is not just evidence in support of the spoiled-milk hypothesis, it is corroborating evidence from two independent sources, the date and the smell. And the evidence itself is interpreted with the help of some background knowledge about the longevity of milk and the link between the sensation (the smell) and the situation (the spoilage). All of this conforms to the essential aspects of scientific method, and all of it happens over and over throughout the day as we come to terms with our environment.

In everyday life as in science, where there is neither supporting evidence, nor logical-mathematical proof, there is no knowledge. And where there is inconsistency in the evidence, or inconsistency between evidence and theory, the responsible thing to do is withhold judgment either way. If the milk smells fine, for example, but the date is a month or more old, common sense suggests an attitude of "I just don't know (yet) whether the milk is spoiled or not." Anything more decisive would be baseless speculation. These are the standards we live by, when we are being honest and responsible, and these are the standards at the heart of the scientific method.

The key difference between science and life on the streets or at the breakfast table is that the scientific process is more deliberate and explicit in following the steps and standards of the method. We do it tacitly day-in and day-out, but in science the procedures must be articulated and described. The scientific process is purposefully slowed down in the interest of control and transparency. A single step of gathering evidence may take days, or weeks, or even years. There are no snap decisions like the ones we are forced to make in crossing the street or preparing breakfast. The components and conduct of the scientific process are essentially similar to those of common sense, but things proceed more slowly and deliberately in science.

Science is also distinguished by a greater dedication to the results of the method than is characteristic of life on the streets. Common sense is, alas, not as common as it should be. Among friends we seem to accommodate such weaknesses as wishful thinking, superstition, and stereotyping. Among scientists we do our best to avoid these violations of the method, following the evidence and the logic as best we can, regardless of our preconceptions.

Science is thus more deliberate and dedicated than non-science in following the method. It is also more public and open to independent review. In life we are rarely asked to actually produce the evidence and reasons for our knowledge claims. We trust each other and ourselves, unless the claim is surprising. But in science, one is routinely expected to not just have good reasons in support of one's claims but to actually produce those reasons. Science is a communal activity in which ideas, procedures, and standards are shared and compared among people. It's out in the open in a way that applications of common sense are not.

All of these features of science, the slow, deliberate, explicit, public application of reasoning from evidence, make the process clear and plainly visible. It's a good specimen to study, a kind of magnified image of good reasoning. And what you learn about scientific method will clarify the details of what we expect of common sense. Reciprocally, an intuitive appreciation of common sense will be helpful in understanding scientific method. There should be no tension between the guidelines of good science and good sense. There should be no reason to suspend the standards of one in the context of the other. You use common sense in the laboratory, just as you would use scientific method of evidence and reasoning during the course of the day, if you took the time to stop and reflect.

Given the continuity between scientific method and common sense, it would be disingenuous to selectively suspend scientific standards in some circumstances or on certain issues. Refusing the authority of evidence and logic, either in the form of believing without evidence or believing in spite of contrary evidence, is not just turning away from science; it is turning away from good sense. Ignoring the standards of scientific method in a particular case amounts to ignoring the standards of good reasoning we all use in every other case. Without an explicit reason to change the rules of evidence and reasoning, it can only be a matter of convenience. It can only be irresponsible.

It is important to restrict this view of scientific method to matters of fact. It does not apply to matters of value. Science, and the empirical side of common sense, can help us figure out the way the world is, not the way the world ought to be. The method of observation, hypothesis, evidence, and testing leads to the knowledge that the milk is spoiled, but gives no support to any claim about whether it is in any sense right or wrong that the milk is spoiled. Science tells us that the earth is in fact billions of years old, but it offers no comment on whether this is good or bad, right or wrong, appropriate or not. Empirical evidence and the method in which it figures do not support value judgments. Scientific method works for description, not evaluation.

Sometimes people ignore this distinction between fact and value, and conclude that things really ought to be a certain way, simply because in fact they are that way. (Mom, everyone is doing it!) This is a logical fallacy. Just because things are this way, it doesn't follow that they should be, or that there is anything should-or-shouldn't about the situation. This reasoning from the facts to an evaluation happens frequently enough that philosophers and logicians give it a special name. They call it the naturalistic fallacy. It is a fallacy, a mistake of reasoning, to argue from "is" to "ought". Even if everyone is doing it, it doesn't follow that everyone ought to be doing it, whatever it is.

Remember this fallacy when it comes to talking about laws in science, laws of nature. Laws, like the law of gravity or the laws of thermodynamics, are more than simply generalizations of what does happen in nature; they are about what must happen in nature. It's no accident that all massive objects attract each other and that things accelerate toward the earth. They don't just happen to do this; they must do this. It's the law. There is a kind of necessity associated with laws of nature. But given the evidence that all massive objects do attract, can we justifiably conclude that they must attract? If we can't imply "ought" from "is", can we imply "must" from "is"?

It is important to be aware of these potential limitations on scientific method. A factual description based on empirical evidence cannot support evaluative conclusions about the way things should be. This would be the naturalistic fallacy. There is also a kind of reversed version of the naturalistic fallacy that should be recognized and avoided. This would be drawing conclusions about the way things are, based on the way you think they should be, inferring "is" from "ought". Wishful thinking (Oh, I'm sure the milk is fine!) commits this fallacy. So does an argument that begins with the assumption that humans are a special and privileged class of beings, and concludes that the earth must be at the center of the universe and stationary.

The reversed version of the naturalistic fallacy has no name. That's a shame, since a label on something raises our awareness. So science will have to get by with an unsung respect for this rule against projecting our own human sensibilities and desires onto nature.

There is another important limitation on scientific method and its companion common sense; neither results in perfect, indubitable certainty. The available evidence and our own abilities to reason are limited. This means there will always be things about nature that we just don't know, and maybe even can't know. It would be irresponsible to fill in these inevitable gaps with speculation dressed up as knowledge. Sometimes, the responsible thing to do is withhold judgment, and scientific method, in its respect for evidence and reason, helps in identifying when that time has come. There is an aspect of wisdom that includes the statement, "We just don't know."

Limited evidence and fallible human reason also result in lingering uncertainty even in what we do know. I know the milk is spoiled, but there is still some small chance my nose is off (human sensations are often inaccurate) and the date is mismarked. But it would be ridiculous to allow this small possibility of error to discredit the argument or the conclusion. This is <u>not</u> a time to suspend judgment. In this case, throw out the milk, not the evidence or the conclusion. We deal with uncertainly like this all the time in life, and we are sensitive to <u>degrees</u> of uncertainty. So too in science.

Any interesting or important case in science will involve some degree of uncertainty, even where the evidence is abundant and the testing unambiguous. Usually science deals in entities or processes that cannot be directly observed, so evidence is indirect and theoretical conclusions are never 100% certain. But this does not mean that theories are pure guesswork. There is an important spectrum of good reason between guesswork and certainty. The challenge of scientific method is to locate a particular theory on that spectrum, and assign our beliefs about nature accordingly. As evidence is collected, a theory can change position on the spectrum. Germ theory, for example, was once very speculative and was much closer to the pure-guesswork end than it is now. Currently, germ theory is so close to the certainty end that you can barely see the separation, but it's there.

This brings us to another fallacy, another mistake in reasoning to expose and avoid. It's the whatever-is-not-100%-certain-is-pure-guesswork fallacy. Listen closely and you will hear it conveniently deployed on a regular basis as a way to clear the ground for someone's pet project. We can't know for sure that x, so we might as well think y. It takes patience to deal with this kind of mistake. It's true, we can't know <u>for sure</u> that x, but we have overwhelming evidence in support of x (and perhaps against y), so the responsible thing to do in almost all cases is to believe x.

There is a good analogy to this fallacy, an analogy that demonstrates the error in if-not-perfect-then-worthless reasoning. Washing your hands before lunch is good policy, because it removes germs that can make you sick. But, of course, it doesn't remove <u>all</u> the germs. And yet it would be silly to say, I'm not washing my hands because it won't get all the germs. You certainly don't want to hear your surgeon say this. It's the same mistake as in saying, I won't believe this theory because the testing has not eliminated all possible doubt. Effective hand-washing comes in degrees, as does effective testing and proving of theories. For hands, cold water is not as good as hot, and soap helps. For scientific testing, it's the degrees of proof and the effective use of evidence that we'll look into next.

If there are degrees of proof and a spectrum between guesswork and certainty, then there must be important details and nuances of scientific method that influence the degrees and locate theories on the spectrum. Understanding these degrees of proof is the most important and challenging aspect of understanding scientific method, so it is to the relevant details that we turn next.

Chapter 2
Empirical Foundations

Abstract All scientific knowledge must be based on observation. This is the basis of scientific method, but there is some ambiguity in how close a link is required between observation and theory. The method cannot be simply a process of generalizing knowledge from observations, since some, at least tentative, knowledge is prerequisite for making scientific observations.

Keywords Theory · Observation · Induction

Clarity on the scientific method begins with clarity on the key terms and concepts. One term in particular, "theory", does a lot of work in most descriptions and evaluations of science, yet equivocation and ambiguity are common. Sometimes "theory" is used in a pejorative way, as an invitation to doubt or even believe the opposite, as in "That's just a theory." But other times "theory" is an honor, implying a coherent network of ideas that successfully explain some otherwise mysterious aspect of nature.

A clear, unambiguous meaning of the term "theory" emerges from a survey of examples of scientific theories and seeing what it is they all have in common, what it is that makes them theoretical. It is not going to have anything to do with how well-tested or well-confirmed an idea is, or how likely it is to be true. Consider some examples: Germ theory, atomic theory, the kinetic theory of heat, the caloric theory of heat, the theory of relativity, the big bang theory, string theory. These have been arranged roughly in descending order of our having reason to believe they are true. The exception is the caloric theory, for which we have good reason to believe it is false. The point is that theories occupy all positions on the spectrum from near-certainty to pretty-speculative, so the term "theoretical" cannot distinguish an idea as being believable or not.

What do all these theories have in common? They all describe objects or events that are not directly observable. This is the core of the concept of theory. A theory describes aspects of nature that are beyond (or beneath) what we can observe, aspects that can be used to explain what we observe. Germs, atoms, caloric, curved spacetime, and elemental strings are all, to one degree or another, unobservable. That's what makes them theoretical. But that doesn't make them unreal.

A theory is true if it describes unobservable things that really exist, and describes them accurately. Otherwise it is false. This shows the mistake in contrasting "theory" and "fact". A fact is an actual state of affairs in nature, and a theory, or any statement for that matter, is true if it matches a fact. Some theories are true (atomic theory), some are false (caloric theory), and the scientific method is what directs us in deciding which are which. To say of some idea, That's a theory not a fact, is a confusion of categories, a comparison of apples and oranges. Facts are; theories describe. And a theory can describe facts.

There is a second term used in describing scientific method that warrants explicit definition, "hypothesis". Unlike "theory" this term does refer to the amount of good reason to believe, that is, to the location on the spectrum between certainty and speculation. A hypothesis is a theory that has little testing and is consequently located near the speculation-end of the spectrum. It's a theory for which the connection to fact is unknown or unclear, but usually there is some tentative reason to believe this link will be made. There is reason to think that evidence and support from other theories will allow the hypothesis to move up the spectrum to the well-supported side.

Being hypothetical is a matter of degree, and the term "hypothetical" wears off gradually. The label is not peeled off all at once. It has pretty well worn off the germ theory, and it is only very faint on the general theory of relativity. It is still distinctly visible on string theory. And again, it is the work of scientific method to determine the appropriate degree of being hypothetical.

One more term should be clarified, "law". Theories differ in terms of their generality. The big bang theory, for example, is about a singular, unique event. It is not general at all, despite being about the entire universe. The theory of gravity, either the Newtonian or relativistic version, is very general. It is about all objects with mass and their resulting attraction. The most general theories, including the theory of gravity, are laws. In other words, laws are theories of a particular kind, the ones that identify whole categories of things and describe their relations in the most general terms. Laws start with the word "all", as in, All this are that, All massive objects are attracted to each other.

Being a law has nothing to do with being well-tested or generally accepted by the community of scientists. A theory is a law because of what it describes, not because of any circumstances of confirmation. And a theory is or is not a law from the beginning, even when it is first proposed, when it is a hypothesis. The status of law is not earned, nor does it rub off; it is inherent in the content of the claim.

So neither "theoretical" nor "law" is about being true or false, or about being well-tested or speculative. "Hypothetical" is about that kind of thing. To describe a statement as a theory, or just a theory, does not imply any implausibility or weakness. A description of gravity, after all, is a theory. Gravity itself is also a fact. The thing about gravity is that it is not just hypothetical. There is a wealth of evidence and good reason to believe in the current theory of gravity. The question now is how, in general, do evidence and reason work to confirm a theory? How does the label of "hypothetical" wear off?

All scientific knowledge must be based on observation. It must have empirical foundations. This is an obvious start to describing the method, but it doesn't make much progress, since the concept of "based on" is vague. It allows different interpretations in terms of just how strictly one specifies the link between observation and theoretical conclusion.

Most conservative is the view that one should endorse claims only when one can directly observe what they are about and observe that they correspond to the facts. In other words, "based on" means explicitly linked to. On this interpretation, theories, since they are about things that cannot be directly checked by observation, can only be regarded as useful models, but never as true descriptions. This is not to say they are false, only that we cannot know which are true and which are false, because in this interpretation knowledge requires direct observation.

But this is just one extreme in the possible interpretations of "based on observation". In our everyday lives, and on most accounts of science, we allow some manner of careful inference that goes beyond the observations. I wake up in the morning and see the ground is wet, the trees are dripping, and there are puddles everywhere. I know it rained, even though I didn't directly observe the rain. And since it is passed, I can't observe the rain. In life as in science, knowledge extends beyond both the observed and the observable. The key is in specifying what kind of inference is reliable for making this extension.

Induction is a prime candidate, and making it the cornerstone of scientific method is the next most conservative account beyond allowing only claims about direct observation. Staying close to the empirical roots would suggest allowing inference that starts with observation and results in theory, not the other way around. It would require a one-way flow of information, from nature to us, from outside-in. The process should be no more than a generalization of what is observed. This is the essence of induction.

Inductive generalization is a very common-sensical process. If you observe that some thing A_1 has a property B, and that A_2 is B, and that A_3 is B, and so on, you eventually conclude that All A's are B. This is bottom-up, outside-in reasoning. The conclusion has come from nowhere other than observations of nature, without fabrication. It is understood that the conclusion is only probably true, but that is no surprise, given our respect for the fact that science does not deliver certainty.

Some people, Isaac Newton among them, have claimed that inductive generalization from observation to theory is all there is to scientific method. Anything else, such as speculative hypothesis, would be an irresponsible first step on a slippery slope to make-believe and mysticism. But inductive generalization can't be the whole story about scientific method, and it is instructive to see the specific reasons why.

First of all, pure induction, with only observations as premises, could never imply a statement about something unobserved or unobservable. How did we ever come up with ideas about germs, atoms, or curved spacetime? It could not have been simply by generalizing on what has been observed, since none of these things has been observed. If the goal of science was simply to catalog empirical generalizations like all metals conduct electricity and all emeralds are green, then pure induction might do the trick. But science routinely does more than that. It offers explanations for how and why metals conduct electricity and emeralds reflect green light. This is the value of science, getting beyond the merely observable, and pure induction does not suffice as the way to do this.

There is a second, and more fundamental, reason why induction cannot be the whole story, or even the most important part of the story, about scientific method. Pure induction presupposes pure observation, an uncontaminated flow of information from outside-in. This is simply impossible. In life as in science, perception is influenced by ideas. Scientific observations are influenced by scientific theories, so the order of events cannot be strictly observation then theory.

It is an important insight into scientific method to show not only <u>that</u> theory influences observation, but exactly <u>how</u> this influence comes about. Here are four reasons why there can be no theory-neutral observations in science.

First, it is impossible to observe everything. Selecting what to observe and what to ignore cannot be haphazard. Science isn't simply a catalog of observations; it's inferences from <u>relevant</u> observations. Some basic idea of what is relevant to what is needed in selecting what observations to make and record. Isaac Newton understood the relevance between a falling apple and the orbit of the moon, but no one else did. No one else had the theoretical predisposition to know what to make of the apple and to know that it should be recorded as an important observation. Without some theory in mind, the falling apple would have been unremarkable and unrecorded.

Furthermore, it is impossible to note and describe every detail of the observations that are made. Selections have to be made, this time regarding the relevant aspects of the observations. And again, the selections are not random or haphazard, but informed by some existing understanding of the situation. It's not the color of the apple that counts, red or golden. It's not the variety, macintosh or delicious. It's not even the shape or size or the fact that it's an apple and not a stone. It's the mass that's relevant, and Newton knew to focus on this property alone. Only some background knowledge would allow him, or any scientist, to safely ignore some details and attend to others.

A third reason for some theoretical imprint on the information from observation is the requirement that scientific observations be careful and reliable. Observing conditions must be proper. Relevant conditions must be controlled. If machines are used, they must be working properly. And so on. Accounting for reliability will call on a theoretical understanding of how the machines work, which conditions are relevant, and what amounts to proper conditions. <u>Scientific</u> observation, unlike casual observation out on the street, is accountable; it is always open to challenges to its accuracy. Meeting the challenge calls on background knowledge. Think of it this way. A novice wouldn't know how to conduct an experiment in chemistry or high-energy physics. They wouldn't know credible evidence from contaminated data, because they lack the expertise to judge good from bad observing techniques. The expertise, what the chemists and physicists have, is based on their theoretical understanding of the situation. That's why they took all those classes in chemistry and physics.

Finally, the novice loose in the lab wouldn't know what to make of what they were observing. Even if the results were selected, highlighted, and certified by experts as reliable, without a theoretical context the description of the results would be useless. "Pink paper" and "streaks in the bubble chamber" must be interpreted into the theoretical language of acid and alpha particle in order to be useful data in suggesting or testing theories. Scientific observations, in other words, must eventually be rendered in theoretical language, and this certainly presupposes a theory in place. To know that the streaks in the chamber mean there is an alpha particle, one must understand the causal chain of events from particle to streak, and this is a matter of theory.

To summarize the ways in which theory influences scientific observation we can say that in science one needs evidence and not merely sensations. Evidence must be meaningful and reliable. It must be a credible indication of something, as the streaks are an indication of an alpha particle. The connection and the credibility are underwritten by theory. Of course the theories used to make the best and the most of the evidence are themselves subject to revision. That's science. But at least some tentative theories must be invoked in the course of scientific observation.

This means that the inductive route from observation to theory cannot be all there is to science. There will have to be a flow of information back-and-forth, from theories to observations and from observations to theories, from inside-out and outside-in. Of course there is an important role for induction, and in some way, ideas for theories come by way of observations of the world. But just as surely, there is an important aspect of the method that comes after the theory has been suggested. Induction plays an important role in the discovery of a new idea, a hypothesis, but there is work still to be done in testing the hypothesis. Sometimes it is suggested that it is the testing stage that is the essence of scientific method. It doesn't really matter how a hypothesis is discovered or how someone first got the idea. Kekulé, apparently, got the idea that the benzene molecule is a closed ring by dozing in front of the fire and hallucinating snakes. So what, as long as the idea is rigorously tested before being entered into the books as scientific knowledge? The important role of empirical evidence may come after the idea is proposed rather than before.

Chapter 3
Empirical Testing

Abstract Empirical testing of a scientific hypothesis is always indirect. A hypothesis is tested by making predictions and seeing if the predictions are true. A look at the logic of this shows that a true prediction cannot prove a hypothesis. Nor can a false prediction disprove a hypothesis. So empirical testing is always indecisive, and scientific method must involve more that just evidence and logic.

Keywords Testing · Confirmation · Falsification

Empirical testing is undeniably an important component of scientific method. Scientific claims must be based on observable evidence, and we have seen that the observations that come before a theory are insufficient basis for claiming the theory is true. So there must be an important role for evidence that comes after the theory is proposed, evidence used to test it.

If a statement is about something that is itself observable, then the empirical testing can be direct. We just have a look to see if it is true. For example, the statement, "The litmus paper is pink," is subject to direct empirical testing.

But science is most interesting and most useful to us when it is describing unobservable things like atoms, germs, black holes, gravity, the process of evolution as it happened in the past, and so on. This is where we will find explanations, and not mere summaries, of what happens in nature. Theories, that is, claims about unobservable things, are not amenable to direct empirical testing, since we cannot just have a look to see if they are true. These claims are nonetheless accountable to empirical testing that is indirect. The nature of this indirect evidence, and the logical relation between evidence and theory, are the crux of scientific method. This is where we have to be the most careful and explicit. A close attention to the details, and a little terminology, will be worth the effort.

Statements about unobservable things can be tested by their observable implications. In other words, to test the truth of a statement x, we reason that "If x is true, then we will observe y." y is an observable implication from x, and it is by

P. Kosso, *A Summary of Scientific Method*, SpringerBriefs in Philosophy, 1,   13
DOI: 10.1007/978-94-007-1614-8_3, © Peter Kosso 2011

observing y that x is indirectly confirmed. If we look for y but don't see it, then x is indirectly disconfirmed (falsified). y is the evidence for (or against) x.

Here is a quick example. When the theory of evolution was first proposed, it had to be empirically tested. In this context, the theory of evolution was the hypothesis. But since the theory describes events that cannot be observed, since they happened long ago in the past, the testing had to be indirect. If the theory is true, then the fossils we observe today should fall (very roughly) into a pattern with older (deeper) fossils being simpler and less diverse than the younger (shallower) fossils. We test the hypothesis by looking at the fossils.

Here is another example. When Einstein first proposed his theory of gravity (the general theory of relativity) it had to be empirically tested. The theory says that gravity is caused by the curvature of space and time, and since neither space nor time can be observed, Einstein had to figure out some observable consequence of their being curved. He predicted that if space and time are curved (as his hypothesis claims), then light rays passing near a massive object (such as the sun) will actually bend. This bending should be observable, and we can test the hypothesis by this implication.

Since we are looking for a general pattern in the logic of indirect empirical testing, it will help to symbolize Einstein's argument. Let **H** stand for the hypothesis in this, or any other, case of empirical testing. Let **p** stand for the implication, that is, the prediction. In the particular case of testing the general theory of relativity,

> H = Space and time are curved.
> p = Light rays will bend when they pass near the sun.

Then Einstein's reasoning is in the form of an if-then statement:
> If H is true, then p will be true.

Or, even more briefly:
> If H then p

This kind of statement, if H then p, is the central premise of indirect empirical testing. Since it is a case of deducing the prediction p from the hypothesis H, any test that involves an if-then statement like this is called hypothetico-deductive testing. The complete test requires observing whether the prediction p is true or not. In Einstein's case, the prediction turned out to be true. To say it this simply, that the prediction was true, is to hide significant complication and uncertainty in collecting the data. But we will put this sort of detail to one side for now, to develop the basic logic of hypothetico-deductive (H-D) testing. This is a case of H-D confirmation. In fact, it was the report of this confirmation, based on measurements of positions of stars seen during an eclipse of the sun, that made Einstein famous. Here was this wild idea, curvature of space, from an obscure and wild-looking guy, that turned out to be true. This was the event that made "Einstein" synonymous with "genius".

The argument can be put in an abbreviated form:

> If H, then p   (this premise is from Einstein's reasoning)
> p_____   (this premise is empirical, from observing stars)
> H

But look closely at the form of this argument. It is not a valid argument. It commits the fallacy of <u>affirming the consequent</u>. The conclusion that the hypothesis is true does not follow from the observation that the prediction was true.

Here is another example of committing this same fallacy, one that transparently shows the mistake involved. Any doctor knows that if someone has malaria they will have a fever. But the evidence of a fever doesn't warrant the conclusion that a patient has malaria. This reasoning would match the abbreviated form of argument just above. It's true that if H (malaria) then p (fever). And let's say it's true of this patient that p (they have a fever). You can't conclude that they have malaria. It could be any number of other fever-causing diseases. This is the fallacy.

So the successful prediction of the bending of starlight did not prove that the hypothesis is true, any more than having a fever proves you have malaria. The hypothesis could be false, as a false hypothesis can make a true prediction. Furthermore, there is surely some other hypothesis, some other description of gravity, that makes the same prediction of bending light made by Einstein's hypothesis. So observing the bending light doesn't tell us which of these alternative hypotheses is true.

The moral of the story is this: a single true prediction does not confirm a hypothesis.

But now suppose the prediction had come out to be false? That is, suppose we had empirically discovered that p is false, that is, not p. This is not what happened in Einstein's case, so we are now talking about an abstract example. This would be a case of hypothetico-deductive disconfirmation. The argument still has the H-D premise. The difference is in the empirical premise.

> If H, then p   the H-D premise
> not p   the empirical premise
> ────────────
> not H

Now the conclusion is that the hypothesis is false. This argument is valid. There is no way that the premises could be true and the conclusion false. The conclusion in this form of argument follows with absolute certainty. Applying this to the case of diagnosing malaria, a patient who does not have a fever cannot possibly have malaria. This is good medical practice, ruling out the possibilities.

It seems as if disconfirmation of a hypothesis can be done with a single test, if the prediction is false. Disproof of a hypothesis, falsification, appears to be decisive in a way that proof is not. This apparent disparity between falsification and confirmation has led many people to claim that the essence of scientific method is falsification. Scientists don't prove theories, according to this account; they try to disprove them. Theories that survive repeated attempts at falsification are the ones to believe.

But this is wrong. Disconfirmation seems so easy and so definitive only because we have ignored many of the important details in the example. In particular, we have ignored the theoretical details of how the prediction was deduced in the first place, and the practical details of how the experiment was done. Filling these in will show that disconfirmation of a hypothesis is no more decisive than confirmation.

We will add the important details by doing another example. As often happens in science, start with a question, a phenomenon that needs to be explained:

What is the source of energy in stars?

Stars, including the sun, pour out enormous amounts of heat, light, and other forms of energy. Where does the energy come from? The most widely held theory about this is that nuclear fusion is the source of energy in stars. That is, light atomic nuclei, like hydrogen and helium, are being joined together to form heavier nuclei like lithium and beryllium. Nuclear fusion releases energy. In fact, it is exactly the same process that is used in nuclear bombs. So the theory is that the stars are exploding nuclear bombs.

The theory states that nuclei of hydrogen are being fused together to form nuclei of helium at the core of the sun where the pressure and temperature are high enough to squeeze the nuclei together. This process is unobservable, because it happens deep inside the sun and it involves subatomic particles. So the theory must be tested indirectly, by figuring out some observable implication of the nuclear fusion.

In this case, the hypothesis to be tested is the theory of solar nuclear fusion:

H = Nuclear fusion is taking place at the core of the sun.

We need experts to predict what observable implications to look for, experts who know the details about nuclear fusion and the byproducts of fusion, about the sun, and so on. One prediction involves an elementary particle called a neutrino. Neutrinos are created by nuclear fusion, and neutrinos are so light and electrically neutral (hence the name neutrino, which means little neutral one) that they will escape from the center of the sun and fly to earth. The prediction then is that we will detect neutrinos here on earth. And the prediction is very precise as to exactly how many neutrinos we will detect, given the amount of energy generated in the sun.

p = A specified number of neutrinos will be detected on the earth.

Astrophysicists went to a lot of trouble to detect these neutrinos and confirm what they were sure was the right theory about the sun's energy. But for decades, no neutrinos were found, at least nowhere near as many as predicted. Yet just about everybody continued to believe that the theory is true. This mismatch between the presumed correct theory and the reliable, but disconfirming, evidence was vexing enough to earn the name "the Solar Neutrino Problem".

The missing neutrinos have recently been found, but the problem had a life of several decades. During this time, were the scientists being unreasonable and dogmatic? Were they simply ignoring the logic? Should the not-p observation have forced them to conclude not-H? No. When we add more of the scientific details, the logic of the H-D argument becomes much more complicated and the conclusion less certain. We will find that in this case, as in all cases of indirect empirical testing, a false prediction does not necessarily falsify the hypothesis.

There are two kinds of details we need to consider, theoretical details of how the prediction was made, and practical details of how the experiment was done. We'll look first at the practical details of the experiment.

Detecting neutrinos is very complicated and tricky. Since they are so light and without electric charge, they don't make much of an impact on other things. They fly right through the sun with little probability of hitting anything. They fly through our bodies with only the rare interaction. They fly through the earth with only a slim chance of getting stopped. It turns out that neutrinos do sometimes interact with the nuclei of chlorine atoms, changing them into argon nuclei. The argon is radioactive, and this is how the neutrinos are detected, by measuring the radio-activity of the argon. So you start with a huge tank of chlorine, huge because it is still the rare neutrino that interacts and you want lots of chlorine targets for them. Then after a while you count the number of radioactive argon atoms.

There is one more important detail, the tank must be shielded from other sources of radioactivity such as cosmic rays. For this reason, the first experiment was done in an abandoned gold mine where the thick layer of earth would block out the background radiation but allow the neutrinos to enter. The experiment consists of a 100,000 gallon tank of cleaning fluid (a cheap source of chlorine) buried 5400 feet deep in an abandoned gold mine. The tank is periodically swept for argon atoms that are detected by their radioactivity. The number of argon atoms is correlated to the number of neutrinos that passed into the tank.

Putting this all together, the prediction p is itself in the form of an if-then statement:

p — If we put a huge tank of cleaning fluid deep under ground and use the
    appropriate radioactivity detector, then we will get a specified number of
    clicks on the detector.

The if-part of this statement is a list of the experimental conditions. These specify how the experiment is to be set up. The then-part is the final expectation, the predicted outcome. Here is a partial list of the conditions:

Experimental Conditions

     C_1 = Do the experiment deep underground.
     C_2 = Use a sufficient amount of cleaning fluid.
     C_3 = Use the appropriate radioactivity detector.
     C_4 = Have the detector warmed up.
     .
     .
     .

There will be lots more experimental conditions, including all the precise details on how to set up the experiment. The experimental conditions are effec-tively the recipe for doing the experiment, similar to what is described in a chemistry or biology teaching-lab book. It is the precision and explicit presentation of experimental conditions that are the key to repeatability in scientific conditions. We need a detailed record of what was done, in case we want to do it again, for ourselves. This is also an essential component in judging the credibility of the data. It specifies just what it takes to do the experiment properly.

The then-part of the prediction is what to expect when the experimental conditions have been properly done. The predicted result of the experiment is the expectation:

Expectation

E = There will be a specified number of clicks on the radioactivity detector.

In general, in abbreviated form, the full prediction p is the if-then statement:
prediction

$$p = \text{If } C_1 \text{ and } C_2 \text{ and } C_3 \text{ and } \ldots, \text{ then E.}$$

In other words, if all the conditions are done right, then we will get the expected results. All scientific experiments are like this, where the final result is carefully controlled by the experimental conditions.

From now on, whenever we use the symbol p to stand for the prediction, we are using it as an abbreviation for the if-then statement involving the experimental conditions and the expectation.

We need to see how the experimental conditions and the expectation affect the logic of H-D testing. But first we'll describe the other complicating factors, the theoretical details involved with making the prediction in the first place. Then we'll put all the pieces together.

It is likely that when you first read the hypothesis about nuclear fusion in the sun you did not immediately think of neutrinos or cleaning fluid. Neither did I. That's because we lack the required background knowledge to deduce the implications from the statement about fusion. It takes an expert, someone who knows about nuclear physics and neutrino physics and the chemistry of chlorine. All that background knowledge is theoretical, in the sense that it is about things that cannot be observed. The individual statements that are drawn from background knowledge and used in deducing the prediction are often called auxiliary theories.

Please note that auxiliary theories are not the ideas used to originally come up with the hypothesis. They are the ideas used, once the hypothesis has already been suggested, to figure out how to test it. Scientists go to school for years and years to learn the basics, and this gives them the background knowledge to deduce observable predictions from hypotheses.

In the case of the solar neutrinos, the auxiliary theories are the statements about the fundamentals of nuclear fusion, production of neutrinos, the flight of neutrinos, interaction between neutrinos and chlorine, and so on:

Auxiliary Theories

> A_1 = Nuclear fusion produces neutrinos.
> A_2 = Neutrinos can fly from the core of the sun to the earth.
> A_3 = Neutrinos interact with chlorine nuclei.
>   .
>   .
>   .

There will be lots more auxiliary theories, describing every step of the process from fusion in the sun to clicks of the detector. And all of this information is required to deduce the prediction. In other words, it's not simply If H then p, it's If H <u>and all the A's</u> then p. The hypothetico-deductive premise is:

$$\text{If H and } A_1 \text{ and } A_2 \text{ and } A_3 \text{ and } \ldots, \text{ then p}$$

And don't forget that p is an abbreviation for the if–then statement involving the C's and E. So the full H-D premise is:

$$\text{If (H and } A_1 \text{ and } A_2 \text{ and } A_3 \text{ and } \ldots), \text{ then (if } C_1 \text{ and } C_2 \text{ and } C_3 \text{ and } \ldots, \text{ then E)}$$

Now what happens if the prediction turns out to be false, that is, the expectation turns out to be false? This is what happened in the solar neutrino case. The <u>empirical premise</u>, that is, the observed result, was: <u>not E</u>. They weren't getting as many radioactivity detections as predicted.

But that did not mean that the hypothesis H is wrong. The problem could be in the experiment itself (that is, one of the experimental conditions isn't done right) or in the background knowledge used to make the prediction (that is, one of the auxiliary theories is wrong).

The argument, in the case where the prediction comes out false, is this:

If (H and A_1 and A_2 and A_3 and ...),

then, (if C_1 and C_2 and C_3 and ..., then E) (H-D premise)

not E (empirical premise)
_____

not(H and A_1 and A_2 and A_3 and ... C_1 and C_2, and C_3 and ...)

The conclusion is that something is wrong somewhere. It can't be that all of the claims, H, the A's, and the C's, are true. At least one of them is false, and that's why the expectation came out wrong. In other words, either H is false, or A_1 is false, or A_2 is false, or..., or C_1 is false, or C_2 is false, or...

It does not follow that it's the hypothesis that's false. It could be an experimental condition that wasn't done right, or it could be a mistake in the background knowledge, a false auxiliary theory. This begins to show why disconfirmation can never be definitive.

There is an important difference between the statements describing the experimental conditions (the C's) and the auxiliary theories (the A's). Experimental conditions are observable. Things described by auxiliary theories are unobservable. This is important because it means that we can directly check the experiment to see if the conditions have all been done properly, that is, to see if the C's are true or not. But we cannot directly check the truth of the auxiliary theories, anymore than we can directly check the truth of the hypothesis.

So if the empirical premise (the statement of the observed outcome of the test) is: not E, we can logically say that H could be true, but one (or more) of the C's or A's is false. Suppose we blame an experimental condition? This happens all the time in science, saying that the experiment was not done perfectly. If you don't get the outcome you expected, check to see that things were set up properly. Make sure things are plugged in and the settings are right. And you can check all this directly, since these things are observable. So if you blamed the unexpected outcome on a flawed experimental condition, you are obligated to check that condition and fix it! You can blame the C's only for as long as it takes to check them and perfect them.

But since you can never directly check the A's (they are theories, after all) you can never be sure if they are true. This is why indirect empirical testing is never definitive, even if the prediction comes out false. It's because one (or more) of the auxiliary hypotheses could be false and could have led to the false prediction. A true hypothesis, coupled with a false auxiliary theory, can make a false prediction. And so, just as a single true prediction does not confirm a hypothesis, it is also the case that a single false prediction does not disconfirm a hypothesis. The reasons are different. A single true prediction doesn't prove because that would be affirming the consequent, like diagnosing malaria on the evidence of fever. A single false prediction doesn't disconfirm because of the complicating role of auxiliary theories.

It comes down to this: Since the auxiliary theories cannot be proven, the hypothesis cannot be disproven.

In the case of the solar neutrinos, the empirical premise was not E. The expected radioactivity-detector clicks were not observed. But nobody concluded that H is false, that fusion is not happening in the sun. They continued to believe H. At first the experimental conditions were blamed, and years were spent calibrating the detectors and checking the tank of cleaning fluid. But this didn't produce the expected results. Then the auxiliary theories were questioned. Perhaps we don't understand neutrinos as well as we thought. Maybe they are produced in the fusion events in the sun's core, but they are trapped in there. Maybe they somehow get steered away from the earth. Maybe there are different kinds of neutrinos, all of which are streaming to the earth but only one of which interacts with chlorine. This last suggestion is the one that eventually saved the day. Revised (auxiliary) theory now says that there are three kinds of neutrinos, but the chlorine detector is only able to capture one kind. It misses two-thirds of the solar neutrinos. Recent experiments, with detectors for all three kinds of neutrinos, have in fact detected the predicted number of neutrinos from the sun.

So where does this leave us in our understanding of scientific method? Empirical testing is never decisive. A false hypothesis can make a true prediction, just as a true hypothesis can make a false prediction. This doesn't make testing worthless. A true prediction is still some indication that the hypothesis might be true, and a false prediction forces us to rethink some aspect of the situation, the hypothesis or the auxiliaries. But this is supposed to be the scientific method, and there must be something more methodical than is suggested by all the vague language of "some" and "might".

The scientific method is based on evidence and logic, but the details of empirical testing show that evidence and logic alone do not settle the issue of which theories are likely to be true and which false. So what else is involved?

Chapter 4
The Network of Knowledge

Abstract Scientific method must involve a broad view of lots of different ideas. A theory is judged by its relation to many different observations and many other theories. Scientific knowledge must be a coherent network of statements, both empirical and theoretical. No piecemeal or isolated view of single theories and single experiments can do justice to the scientific method.

Keywords Coherence · Expertise

No single test, no single piece of evidence, can confirm or disconfirm a hypothesis. This is probably no surprise, at least not the confirmation part. But it is something to keep clearly in mind when considering what the scientific method can do and when evaluating the plausibility of particular theories. Remember that in science, the request to, "Give me one good reason to believe ..." is simply unreasonable. It misses a crucial point about confirmation in science. Reasons to believe (or disbelieve) can only be good when they come in numbers greater than one.

But empirical testing isn't simply a matter of numbers of confirming or disconfirming predictions. Surely some tests are more important, more conclusive, than others. Scientists exercise discretion in valuing some evidence over other. And while a greater number of tests is good, a greater variety of tests is even better. In other words, the number of confirming or disconfirming data is not as important as a disparate variety of data. Lots of positive predictions do not avoid the fallacy of affirming the consequent any better than does a single positive prediction, if the lots are essentially a repetition of a single kind. But often, what counts as a relevant variety is a matter to be judged by an expert. Newton, for example, knew that testing his hypothesis about gravity on several varieties of apples would be a waste of time. Even dropping different fruits, or fruits and vegetables, would not add to the empirical confirmation of the hypothesis, as these would be the same experiment over and over. The most important kind of difference, the broadest evidential base, came in testing the hypothesis in both terrestrial and celestial contexts. It worked for apples, and it worked for the moon.

P. Kosso, *A Summary of Scientific Method*, SpringerBriefs in Philosophy, 1,
DOI: 10.1007/978-94-007-1614-8_4, © Peter Kosso 2011

Some tests are more important than others, and variety of evidence is an important aspect of empirical testing. Again, this is probably no surprise. But notice how the words "discretion" and "judgment" have appeared in the description of scientific method. These are suggestive of a subjective component to the process. Discretion apparently defies methodical rules, and judgment invokes an ineffable influence of other things one might have in mind. Maybe some important aspects of science depend on those unregulated, unarticulated actions of the human mind, things like intuition, artistry, finesse, and insight. But then we seem to be on our way to those patently unscientific human actions like prejudice, faith, and dogma.

The point, or actually the question, is that scientific method cannot be simply a matter of objective information in observation and logic. It must be observation and logic, and something else. Is the something else a subjective, and hence personal and idiosyncratic, contribution of the scientists? The short answer is no, at least not to the extent that the reliability of scientific knowledge is significantly compromised.

We have already seen that scientific observations are not the pure flow of information from the outside world that a naïve account would suggest. Observations are interpreted and accredited by theories, and theories are devised by human scientists. So there is this contribution from the mind of the scientists, but it is not subjective in the sense of being personal. The influential theories are not a matter of whim or taste.

But there is still a possibility of circularity in the method, a logical weakness that is no less compromising than unchecked subjectivity. If theories are allowed to rule on what the evidence means and when it is admissible, and if that evidence is then cited as reason to believe the theories, it's a tidy insularity that seems guaranteed to result in confirmation. But that's not what happens, at least not when the method is correctly applied. The key to avoiding the circularity is independence. Any theories used to influence the observations must be independent from the theory being tested by the observations, the hypothesis. Consider the case of the solar neutrinos. The hypothesis is about astrophysics and the energy in the sun and other stars. The auxiliary theories are about the chemistry of chlorine and the fundamentals of elementary-particle physics. The theories used to influence the interpretation of the observation come from different books. There is no self-serving situation of a theory sponsoring its own evidence.

The point here is that the scientific method, and the information gained through observation, can be essentially under the influence of what the scientists have in mind, without compromising the objectivity of the method or the information. The accomplishment of confirmation is achieved to the degree there is coherence between theoretical and empirical information. Each influences the other in a reciprocal way, and the reason to believe the results is not in any single, local agreement. The reason to believe any of it can only be seen in a broad view of widespread coherence among ideas.

The discretion and judgment that were called on earlier to decide questions of importance of data and relevance of variety are not matters of taste or whim or prejudice. They do not appeal to that kind of personal subjectivity. They appeal to the authority of existing scientific theories and the requirement to maintain overall coherence in the system of beliefs. It's not personal influence over the judgments; it's theoretical influence. That's why the discretion is not in the hands of just anyone off the street. It's in the hand of the experts, since their expertise gives them the theoretical context to make theoretically-informed decisions.

The good reason to believe a scientific claim is a matter of overall fit into a coherent network of other claims, theoretical and empirical. This means that no one can responsibly judge the plausibility of a scientific claim quickly or in isolation from its broad context. There is no short course to scientific expertise, no short cut to making informed decisions. I have to admit that I am in no position to judge for myself whether string theory is likely to be true. I have to leave this up to those who have spent the time learning the whole story. For the same reason, school children are in no position to judge for themselves whether the theory of evolution is likely to be true. Scientific judgment is informed judgment, and that requires knowing a lot of the context.

These same standards of judging an idea in its context and evaluating its plausibility by global fit apply in our everyday lives as well as in science. In life, simple, direct observations usually rule the day without explicit interpretation or accreditation by our background knowledge. This is because our everyday observations are, well, everyday. We repeat the same circumstances over and over, and we implicitly know what the appearances mean and when they are reliable. There are few new or challenging situations in life, as there frequently are in science. But even in life, observations are sometimes overruled or reinterpreted by the mind. This happens when what we see, or think we see, clashes dramatically with what we already understand. The Gallatin River in Southern Montana looks in places to be flowing uphill. But that just can't be. To allow that observation into the system would conflict with the basic understanding of gravity and liquids. It would compromise the coherence in the system. So the observation is discredited, demoted to the status of optical illusion.

There is an important way in which the structure of everyday knowledge differs from the structure of scientific knowledge. In the everyday, we all share a common network of basic background beliefs. We all have the expertise required to put ideas in context, and we do it implicitly and without really thinking about it. So we can all make informed judgments about the everyday ideas. We are our own well-qualified experts in life, but not in science.

The concept of coherence in a network of ideas is playing an important role in this account of scientific method, and it deserves as much precision as possible. The most basic requirement of coherence is logical consistency. A network of scientific knowledge cannot tolerate contradiction. This is not to say that there are no contradictions lingering in the sciences, but where they are identified they must be addressed. Inconsistency between theory and observation plagued the solar neutrino problem, and that's why it was such an important problem and an area of intense work. Contradiction cannot be ignored.

Scientific claims must not only be consistent, they must be cooperative. This is less precise than logical consistency, but it requires not just compatibility in the network of ideas, but connections among the ideas. Theoretical claims explain observations, and sometimes they explain other theoretical claims. One theory participates in the role of auxiliary in accounting for the evidence of another theory. And so on. There is a variety of kinds of links between scientific ideas. And building such an inter-related, coherent web of claims is a challenge and an accomplishment.

Scientific publications often have many more than one author. This reveals something about the structure of scientific knowledge, namely, the necessarily web-like connections of ideas. Many authors can bring many areas of expertise and thereby link many ideas together. This is what scientific testing and scientific reasoning are all about. When disparate, independent sources of information converge to a consistent, cooperative conclusion, there is good reason to believe the conclusion is true. The more tangled the web, tangled in the sense of far-reaching and inter-connected, the more likely it is accurate. Weaving a tangled web in science, where some of the components must be matters of observation, would be difficult indeed, unless it was on to the truth.

It is worth repeating the big idea at this point, because it is often over-looked in descriptions of scientific method, even though it is an essential aspect of scientific credibility. Scientific method cannot be piecemeal, testing one idea by appeal to another, testing one hypothesis using one observation. The method must be broader than that. It must consider the broader network of ideas. Furthermore, science is not a process of settling one issue and moving on to the next, accumulating truths. There is an ongoing back-and-forth among ideas, between theory and observation, and between theory and theory. Entries in this dynamic web of knowledge are never written in pen. Science is recorded in pencil.

The classification of being hypothetical wears off slowly. It wears off as the theory is linked to more and more observations and to more and more other theories. As the hypothesis comes into equilibrium within the network of scientific knowledge, there is more and more good reason to believe it is true. This is the scientific method.

Chapter 5
Scientific Change

Abstract The history of science shows significant changes in theories describing nature, and scientists themselves will admit that their discipline is dynamic. If things keep changing, what reason is there to believe what's on the books at the moment? The scientific method must include some persistent mechanism that allows for change but in a way that, at least in general, increases the chances that the results are true.

Keywords Paradigm

Scientific theories require evidence. Descriptions of nature are answerable to observations of nature. Similarly, descriptions of science are answerable to observations of science itself. Our account of scientific method requires evidence, examples of real science in action. We should compare what we've said of science to a variety of cases, and not just current cases. Using only contemporary examples would be too narrow. A variety of evidence makes for more meaningful testing, so a variety of cases from the history of science is called for. This means long-term history of science, not just to the beginning of the solar neutrino problem, or even the introduction of Einstein's theory of relativity. We should apply the model of scientific method to Darwin's introduction of the theory of evolution, to Newton, Galileo, and even Aristotle.

As in science, where the essence of the method is dynamic give and take between theory and observation, a look at the history of science should allow for give and take between the model of the method and the description of the cases. What we see in the examples will be interpreted under the influence of our understanding of how science works. At the same time, that understanding must be amenable to some changes if the examples just aren't making sense. Science itself is neither bottom-up nor top-down. Neither is the history of science. Both are meet-in-the-middle, between theoretical and empirical information.

The history of science is the history of some important and fundamental changes. Scientists themselves will tell you how dynamic their disciplines are. You

P. Kosso, *A Summary of Scientific Method*, SpringerBriefs in Philosophy, 1, DOI: 10.1007/978-94-007-1614-8_5, © Peter Kosso 2011

have to be on your toes in science. Textbooks go through new editions quickly, and relearning even some of the basics is part of profession. No one retires with the same network of knowledge they learned in school.

But the changes during a career are usually changes of details. More important and challenging to our understanding of scientific method are the profound changes that have occurred over the longer course of science. Here is a recent but important change that will directly affect your health, if you ever have an ulcer. It used to be common knowledge that a stomach ulcer was caused by peptic acid, often a result of a stressful lifestyle or spicy foods. But that's just not true. We now know that ulcers are caused by bacteria, and they can be treated with antibiotics. This came as quite a surprise, as medical scientists had thought that no bacteria could survive the normally acidic environment of the stomach. It has taken some extensive rethinking and revision in that area of the network of knowledge to accommodate the new idea of a bacterial cause of ulcers.

There are, of course, even bigger and more profound changes in the history of science, revolutionary changes. Darwin's introduction of the theory of evolution required a wholesale shift in thinking about nature, from a place of purposeful, intentional events to one of random changes and natural selection. In geology, tectonic plate theory and continental drift required a new understanding of the earth as a planet of gradual, rather than abrupt and catastrophic changes. The kinetic theory of heat replaced the caloric theory and made heat a property and a process rather than a substance. The Copernican Revolution, or, as Galileo put it, the new world system, replaced the old world system of Aristotle and Ptolemy and, among other things, relocated the center of the universe. It also revealed that the earth, despite all obvious observations, does not stand still. It both revolves around the sun and spins on its axis, really fast.

In the first half of the twentieth century, with the introduction of quantum mechanics and the theory of relativity, the mechanical world of Newtonian physics was turned on its head. As in the previous cases of revolutionary science, the changes involved were profound. The new physics is not simply minor adjustments to the old. Nor is it a matter of restricting Newtonian physics to a limited range of circumstances, big things and slow things. Quantum mechanics and relativity show that Newtonian physics is false. It got some basic ideas wrong, ideas like the nature of space and time and causality. Newtonian physics uses its misconceptions, that space and time are independent and absolute, and that all events in nature are deterministic, to generate accurate descriptions of phenomena within the limited circumstances, the big and the slow. But then a theory that says the earth is flat gives you an accurate description in the limited circumstances of your backyard or even your whole state. But it's false, really false. The shift from Newtonian to relativistic and quantum physics is like the shift from flat to round earth. It entails that the previously held beliefs were profoundly mistaken.

This brief list of important changes in the history of science is cause for concern. It shows that fundamental ideas, well tested and well entrenched in the network of scientific knowledge, come and go. So what good reason is there for believing that the fundamental ideas currently on the books will stay? What good reason is there for believing they are true?

The description of scientific method in terms of an interwoven network of ideas might actually make this concern more worrisome. Thomas Kuhn, the man who introduced the words "paradigm" and "paradigm shift" into the general academic and cultural vocabulary, picked up on this. Kuhn describes science in the following way.

A mature science is usually under the influence of a well-entrenched body of background knowledge. These are the ideas found in the textbooks, the kinds of things you have to know to be a scientist. These are the fundamentals and are shared without disagreement and without challenge or doubt. These fundamental ideas provide guidelines on how that science is to be done. The background knowledge is influential in prescribing what experiments are worth doing, how they are to be done, and how to evaluate and interpret the results. The background knowledge also establishes the language, the technical terms, used to describe both theoretical and empirical details. And the background knowledge enforces what Kuhn describes as the basic metaphysics of the science, the broadest kinds of assumptions and categories for describing nature. Aristotelean physicists and astronomers, for example, divided nature into two independent realms with distinct kinds of laws, the celestial and terrestrial. The Copernican Revolution, made precise with Newtonian mechanics, ignored this separation and applied one set of universal laws to both the heavens and earth. This is a profound metaphysical change.

The network of background knowledge, descriptive categories, technical language, experimental guidelines, and so on, constitute what Kuhn calls a paradigm. It is a "disciplinary matrix", a "network of commitments". This last way of defining a paradigm is very telling. It is a network, indicating an interdependent connection among all the components. You can't change a small part of a paradigm without changing pretty much the whole thing. It is all thoroughly laced through with the fundamental ideas of the background knowledge and metaphysics. A paradigm is an unbreakable whole. And it is a network of commitments, indicating that it is formed of ideas, practices, and standards that the working scientists accept as true. There is nothing hypothetical about the components of the paradigm. These are the ideas the scientists use day in and day out.

According to Kuhn, there can be no science without a paradigm. There must be this stable, authoritative body of commitments for scientists to communicate and do more than start from scratch each morning.

On very rare occasions, a scientific paradigm will change. Given the tight network, unbreakable-whole, nature of a paradigm, the change cannot be piecemeal, partial, or gradual. A paradigm shift must be wholesale and abrupt. It must be a scientific revolution. Kuhn's motivation for saying this is based on his survey of historical cases, but it could just as well be argued in principle, based on the essentially holistic nature of paradigms. The influence of a paradigm is pervasive in that it establishes the appropriate language of the science, the background knowledge against which new ideas are compared, and even the standards by which new experimental results are judged. All aspects of a science are thus paradigm-dependent, including any guidelines for judging the acceptability of a paradigm. In other words, paradigms underwrite their own acceptability.

Change from one paradigm to another, according to Kuhn, is not governed by any rational, methodical standards. Since each paradigm judges itself by its own standards, there are no external, paradigm-independent guidelines for determining that one paradigm is better, more likely to be true, than another. During a revolution, a paradigm shift, the change cannot be measured as progress, since no standard of measurement is preserved from one paradigm to the next. We of course judge old paradigms as less accurate than our own. But this, according to Kuhn, is an inappropriate application of the standards of one paradigm to the evaluation of another. Given the holistic, pervasive nature of a paradigm, each must be judged by its own standards.

Kuhn's model of science and scientific change must be taken seriously. But it doesn't force us to renounce the rationality or objectivity of science. For one thing, it relies on a description of scientific evidence as being thoroughly dependent on the authoritative paradigm. It implies that the evidence itself, grounded in observation, supplies no external, objective constraints on theorizing and on the paradigm itself. Presumably, observations can always be re-interpreted, disregarded as unimportant, or discarded as badly collected data if they are inconsistent with the core ideas of the paradigm. The Newtonian paradigm, for example, had trouble explaining why all the planets orbited the sun in the same direction, a clear observation that Descartes' theory of gravity handled quite naturally. Descartes explained gravity as a vortex of universally pervasive ethereal fluid, pulling planets toward the sun like corks circling the drain. So Newton declared the planetary fact unimportant, unworthy of explanation, just one of those things like why there are seven planets instead of twenty. With this authority to rule on what does and does not warrant explanation, a paradigm seems to be insulated against any empirical findings against it.

But this is overstating the effectiveness of Newton's or any other paradigm. Theories and the paradigms in which they function certainly influence observation, but they don't determine them entirely. The difference is very important. Kuhn's model is completely inside-out, not in the sense that he has science backward, but in the sense that it describes the flow of information as one-directional, from theories and paradigms to the reports of evidence. It ignores the essentially reciprocal relation between observation and theory, that each influences the other, and the flow of information is back and forth. The truth is that, under the influence of a paradigm, <u>some</u> evidence can be ignored, re-interpreted, or explained away, but not all of it. This leaves the observations, the information from nature itself, quite a bit of independent authority to which theories must bend.

Some observations simply cannot be ignored or denied or interpreted away, and these serve as paradigm-independent good reasons to believe one paradigm over another. Once it was discovered that Venus went through phases, similar to the moon, the geocentric model of the universe lost grounds for belief to the heliocentric. Once it was shown that ulcers could be cured by antibiotics, it was rational to believe that ulcers were caused by bacteria. And so on.

You get to ignore some of the data, but you can't ignore all of it, or even a significant part of it. The standard that any paradigm must meet is to increase the extent and coherence of the network of knowledge, and to do this such that the ongoing observations, the information from outside-in, fits into the network. This standard of evaluation is not paradigm dependent, and it is the primary standard by which to evaluate good reason to believe scientific claims.

It is also important to note that our ability to observe nature expands over time. This in itself gives later paradigms an advantage over earlier. Tycho Brahe tested the new, heliocentric world system by looking for parallax motion in stars. Finding none, he concluded that the Copernican model must be false. But he did this before the use of astronomical telescopes. Subsequent measurements, made more precise with the use of telescopes, showed that the parallax motion did happen, as predicted by the new world system. The refined ability to observe nature allowed for this new, pivotal information.

The case of stellar parallax as evidence against, or for, the hypothesis that the earth orbits the sun is a good place to demonstrate the important aspects of empirical testing developed in Chap. 3. The hypothesis that the earth moves is indeed about a phenomenon that could not be directly observed, at least not at the time of the Copernican Revolution. It had to be tested indirectly. As early as Aristotle, parallax motion of the stars, the apparent change in angular position due to the changing perspective of the observer, was suggested as an observable consequence of the hypothesis. There are explicit experimental conditions. Measure the angular separation between two identifiable stars, and then measure the separation several months later. The expectation is that the angle will change over the course of months. There is an implicit auxiliary theory at work, as there always is. In this case it is that the stars are close enough to the earth to make the parallax detectable. Simple optics, or even careful observation, makes clear that the effect of parallax decreases as the object is more distant from the viewer. If the stars are too far away, the effect may be there but too small to measure with the available technology. Copernicans could appeal to this and locate the doubt in the auxiliary theory rather than the hypothesis as a way to avoid the disproof of the motion of the earth. It was not until 1838 that Friedrich Bessel and improved technology with the telescope were able to detect stellar parallax.

It is sometimes an interesting and controversial matter as to what counts as an observation and whether the information allegedly observed is independent of the paradigm. For example, sophisticated particle detectors like bubble chambers show the tracks of electrons and alpha particles. Is this observing the particles? Or is this just an elaborate way to use the current particle paradigm to turn a theory into a visual image? We see images of DNA, as another example, produced by an ingenious process of gel electrophoresis. But if you had doubts about the existence of genes or the accuracy of the basics of genetic theory, doubts, in other words, about the accuracy of the paradigm, would this image be convincing? The answer to the question, How do I know this is DNA and these features are genes? will be in terms of genetics and DNA.

But there are plenty of cases where new and improved observation can be licensed in a paradigm-independent way. If you have doubts about the accuracy of the astronomical image you see in a telescope, look through the telescope at some object on the earth and then walk over for a naked-eye look to see that the object matches the image. Microscopes are a little trickier, but there are borderline specimens, just barely visible by both naked-eye and microscope. Microscopes can be checked by photographically shrinking a test pattern and then microscopically magnifying it. The point is that science progresses, and the progress can be judged in a paradigm independent way, in part by improvements in our ability to observe.

Kuhn's model of science may over-state the influence of a paradigm, but it is importantly humbling. We can't ignore the influence of our own background knowledge in evaluating the plausibility of new, or old, ideas. And we can't ignore the uncertainty of current scientific beliefs. There is always some chance that what is in the textbooks now will be replaced in a dramatic and profound paradigm shift. But uncertain doesn't mean false, and it doesn't mean without good reason to believe. You should still wash your hands before lunch.

Chapter 6
Scientific Understanding

The business of science involves more than the mere assembly of facts: it demands also intellectual architecture and construction.

Stephen Toulmin

Abstract Descriptions of scientific method usually focus on the production of knowledge while overlooking what amounts to understanding nature. The holistic account remedies this oversight. Understanding requires apprehending the connections between theories and the way things fit together in the network of knowledge.

Keywords Beauty · Necessity

Of course Toulmin is right, but you couldn't tell this from most renditions of scientific method. They are usually more concerned with the individual bricks than with architecture, dwelling on the testing of each individual theory without much regard for how theories fit together.

Respect for the intellectual architecture does show up, however, in science teachers' attitudes about exams and testing. The difference between knowledge and understanding, and the recognition that understanding is a valuable achievement beyond mere knowledge, is demonstrated by contrasting ways of testing students. True/false or multiple-choice exams test for knowledge, but not for understanding. The answer to each question could be known in isolation, and a perfect score is possible without any idea how one answer relates to another. This kind of test shows what the students know, but not what they understand. Testing for understanding requires longer answers in which ideas must cooperate. Deriving one result from another, or applying ideas to novel situations demonstrates understanding. This is why most teachers value discursive, problem-solving exams over the true/false format. It is also why a lonely result with no derivation, even a correct result, is not a satisfactory answer when asked to solve a problem. "Show your work," means show that you not only know the right answer, but that you genuinely understand what is going on. It means to articulate the links between ideas and clarify the structure of the theoretical neighborhood where this idea lives. It means that the coherence among facts is a necessary part of the answer itself, and not just the means to finding the answer.

P. Kosso, *A Summary of Scientific Method*, SpringerBriefs in Philosophy, 1, DOI: 10.1007/978-94-007-1614-8_6, © Peter Kosso 2011

The intellectual architecture that Toulmin talks about is found in the structure of relations among theories. This is what is needed to not merely <u>know</u> about nature, but to <u>understand</u> it as well.

Another way to think about scientific understanding is in terms of Pierre Duhem's caricature of two kinds of scientist. Duhem was a physicist and philosopher involved in the heady times of physics in the early part of the twentieth century. Some scientists have what he called strong and deep minds, given to focusing on a single issue and pursuing it to its foundations. Others approach science with ample and shallow minds, driven by empiricism and an insatiable appetite for a diversity of facts.

Duhem despaired that neither of these styles of science achieves understanding. The strong and deep scientist tries in vain, while the ample and shallow doesn't try at all. The deep approach fails because it is too narrowly focused, too localized, and understanding is an essentially global achievement. The disciplined focus of the deep mind is blind to the far-reaching theoretical context that is the content of understanding. By contrast, the ample-minded scientist doesn't even attempt to understand what she knows, because her knowledge is too episodic. Many individual facts are described, but an unwillingness to think beyond these empirical basics precludes any attention to how the pieces fit together.

Here is a metaphor about mosaics that highlights the differences between Duhem's two approaches to science, and that clarifies the deficiency in each. The deep-minded scientist is a specialist, an expert on a single tile of the mosaic. He knows in profound detail the shape, color, composition, and properties the layman never knew existed. The ample-minded scientist, by contrast, collects and describes lots of individual tiles, but he pays no attention to how they fit together. He constructs a taxonomy of kinds of tiles, labeled in a language unfamiliar to the rest of us. He may even curate a museum of tiles, individually displayed and grouped in rooms according to kind. Both of these scientists produce a kind of knowledge, as long as they follow standards of testing their claims against what they observe in the tiles. But they both miss the point. Neither approach makes sense of the big picture, and, in an important way, no small piece makes sense except in its context of the big picture. This is true in science as well.

As the mosaic metaphor suggests, there is an aesthetic aspect to scientific understanding, and a sense in which a failure to understand is a failure to grasp a kind of beauty in science. Certainly neither kind of mosaic-tile expert really appreciates the beauty of the mosaic itself. The beauty is in the overall array, in the way the pieces fit together, not in any local or isolated piece. More generally, there is a kind of beauty in the arts, from literature to music to painting, that can be characterized in terms of proper fit and balance among the elements. The beauty in a musical masterpiece is surely not in any single note but in the way the notes cooperate. When no note is out of place, and we apprehend a natural, sometimes inevitable pattern. Sensing the beauty is sensing what will come next, indeed, what must come next. In this way it is like the ability to derive scientific results. The listener or reader or viewer often does not know explicitly what piece goes where or how to go on from here, but they are the laymen in this situation. Experts, those with an artistic sense, have the larger view and a grasp of the essential structure. Laymen are able to tell when things go wrong, when notes are out of place or the character of a novel does something horribly uncharacteristic, or the lines or colors of a painting just don't belong together. For the artist or for the audience, this kind of beauty is a matter of fit in the pattern. In this way, scientific understanding is related to beauty. Both are achieved only in the careful consideration of relations and coherence among the pieces.

Steven Weinberg, a Nobel Prize winning physicist, offers a helpful character-ization of beauty in physics that is similarly linked to our sense of understanding. He promotes the epistemic value of "the rigidity of physical theories". Theories are rigid to the extent that their pieces fit together in a way that no small detail can be changed without large-scale disruption in the coherent network. There is a kind of inevitability and necessity in a theoretically rigid description of nature, in the sense that the values of parameters and the structure of interactions <u>must</u> be as the theories describe them, in order to maintain consistency and connections in nature. Just as no individual note can be changed without diminishing the beauty of a fine piece of music, no theoretical claim can be so disconnected as to be alterable without consequence.

As an example of theoretical rigidity, Weinberg cites the fact that the force of gravity decreases as the inverse square of the distance between two objects. In the context of Newtonian theory, the inverse-square relation is empirically motivated. It is put in the field equations to accommodate observations of planetary orbits. It could have been an inverse-cube, if that was what was needed to save the phe-nomena. But in the general theory of relativity, the inverse-square is the only possible relation. "[G]eneral relativity <u>requires</u> that the force must fall off according to an inverse-square law." There is a theoretical rigidity, a secure fit among ideas in the general theory of relativity. We know that the force decreases as the inverse-square, and we know this in both Newtonian and relativistic con-texts. But only in the general theory of relativity do we see how this fact nests in the network of theories, and only with the general theory of relativity can we claim any understanding of the relation.

Articulating the structure of understanding helps to clarify the comparison between natural and social sciences. This comparison is usually made with an emphasis on the differences between the two domains of sciences, differences such as the precision of testing or the universality of laws. The issue is generally more explicitly debated on the social side than on the natural. Particularly in history and archaeology, there are methodological questions of how like the natural sciences these disciplines are or should be. The distinction is more-or-less assumed by scientists on the natural-science side, but this is in part a result of overlooking the role of scientific understanding. The structure of understanding is in fact an important point of intersection in the methods of social and natural science.

There is no doubt that social and natural sciences study fundamentally different kinds of things, but this does not imply that all of the important aspects of their methods must be different. The social sciences describe human beings, things that think and feel and react with emotion. It is this mental component that makes the specimen different in kind from what is studied by the natural sciences. The added complication of idiosyncratic behavior, free will and agency, makes the metaphysics of social science distinct from natural science. But the difference of the object does not necessarily force a complete difference of method, and it is worth looking for and clarifying any methodological similarities.

Understanding is more commonly associated with social sciences than natural sciences. We may know in a detached way what other people or cultures do, but a genuine connection is made only when we can understand their actions. Knowledge without understanding is clearly an inadequate connection to other people. But how do we understand other people? An account of understanding in social science will shed light on the role of understanding in natural sciences, and hence of this similarity between the two kinds of science.

Understanding in social science is often portrayed in terms of hermeneutics, a method that is explicitly holistic and global. Individual ideas, actions, or symbols are meaningful only when they can be fit into their larger context. The big picture gives meaning to its own individual components. Ideas in isolation are without meaning, as individual mosaic tiles are without aesthetic value. Understanding, like mosaic beauty, is entirely a matter of fitting into a pattern. Understanding depends on coherence.

A good example of the hermeneutic method is R.G. Collingwood's description of historical knowledge. The only way to know the human past, according to Collingwood, is to rethink the thoughts of the people in the past, and to do it under the influence of the appropriate historical context, namely their historical context. This is not a purely subjective empathy but a methodical and regulated "critical thinking". The past is re-enacted and rethought under strict guidelines of coherence. Each datum must make sense in context. It must fit coherently into the emerging picture, as the mosaic tile must contribute to a discernable image. Individual actions must make sense to us, but to us immersed in the context of the past. In this way, the actions must make sense to the actors at the time, and it is this coherence that allows us to say we understand the actors. It's not the individual contents of their thoughts that are the basic data of our understanding them; it's the structure of their thoughts, how things are connected.

Understanding human beings in this sense is recognizing how their individual actions and expressions fit together. The broader the fit, that is, the more far-reaching the coherence among actions and other events in context, the better the understanding. It's in the connections and patterns among our theorized descriptions of other people, past or present, that we are able to understand them. This understanding amounts to knowing how to continue from one idea in context to another, and the ability to apply new events, new data, to old contexts. Understanding in social science is the recognition of the intellectual architecture.

Understanding in natural science follows the same standards of conceptual coherence. It requires attention to relations among theoretical claims, seeing how each one fits into the web with others. Awareness of the links between theories is what allows for the application of ideas to new circumstances. It is the basis of being able to derive values in natural science, without having to measure them. This is a hallmark of understanding in both the natural and social sciences.

Though Duhem associated the strong, deep minds with French scientists, and the ample minds with English, it was a Frenchman who created the finest caricature of the ample and shallow minded scientist. The Omniscienter, a character in René Daumal's novel *A Night of Serious Drinking*, boasts (or laments) on a sign over his chair, "I know everything, but I don't understand any of it".

How could this have happened? I suspect the Omniscienter has spent too much time gathering evidence and too little time thinking about it. He has taken the idea of empirical testing too seriously and overwhelmed his science with observation. Too much data has left too little room for understanding.

There are examples of knowledge without understanding in the physical sciences, and they are found in the most empirically dependent sciences or in any science at the time of new empirical discovery.

A case of an empirical generalization that falls short of understanding is Bode's law. Discovered in the eighteenth century, Bode's law is a mathematical formula that fit the orbital radii of the six known planets into a numerical pattern. Even Uranus, the seventh planet, discovered after the law was written, fit the pattern. But Bode's law had no connection to any other properties or phenomena in planets or in anything else in nature. It was an isolated empirical generalization. Nothing, not the theory of gravity nor theories describing the origins of planets, could be linked to Bode's law. The formula could have been quite different, with no effect on other theories. It is this lack of theoretical coherence and rigidity that supports our intuition in saying that Bode's law described the facts of planetary distances but offered no understanding of the facts.

Another example is quantum mechanics. Weinberg cites the general theory of relativity as the best example of the theoretical rigidity that is the hallmark of understanding. He has less praise for the theoretical rigidity of quantum mechanics, and, not coincidentally, quantum mechanics offers less understanding than does the general theory of relativity. The Copenhagen interpretation, the most common attitude among those who do quantum mechanics, is roughly an abdication of understanding. Quantum mechanics works; just leave it at that. An accurate description of nature requires such uniquely quantum mechanical phenomena as superposition states and non-locality; this we know. But we have no real understanding of such things, since they are incompatible with other concepts and theories we hold. Quantum mechanics doesn't fit into the rest of our conceptual system. We <u>know</u> about non-local correlations of spin-orientations in the EPR experiment, but we don't <u>understand</u> them, and that is because, try as we might, we are unable to link quantum mechanics with other theories and familiar concepts. The theory is coherent in itself, but not in a larger conceptual context with other theories.

There is thus a distinction between quantum mechanics and the general theory of relativity in terms of understanding; there is also an empirical distinction between the two theories. Quantum mechanics is empirically driven in a way that the general theory of relativity is not. The quantum theory was put together to accommodate a series of perplexing experiments such as the photo-electric effect and spectral measurements of black-body radiation. And quantum mechanics is famous for its experimental successes, both in being right on all experimental predictions and for having numerous technological applications. The general theory of relativity, on the other hand, is more theoretically driven. It was developed out of respect for some fundamental principles, such as the principle of relativity, the principle of equivalence, and the principle of general covariance. Its evidential basis, while positive, is comparatively small. So, while quantum mechanics benefits from a firmer empirical foundation, it offers less in the way of understanding. Observation is a fine basis for knowledge, but not for understanding, because observation accommodates piecemeal, isolated data without regard for their structural coherence.

Focusing too finely on the link between theory and evidence leaves us blind to the network of links between theory and theory. Emphasizing the justification of individual claims to knowledge distracts attention from the inter-theoretic relations that constitute understanding, and the resulting strict empiricism leads to the poverty that is knowledge without understanding.

Empiricism leaves us without understanding. It also leaves us, according to the eighteenth century philosophy David Hume, without knowledge of laws of nature. These two limitations are in fact the same thing, since understanding amounts to the apprehension of necessity. To make this clear, consider these two examples of things we know about the natural world:

No causal signal is faster than light.

No animal runs faster than a cheetah.

The first describes a general fact that is of central and fundamental importance in the structure of nature. The second describes a singular, incidental property of one component of nature. The first is a law of nature; the second is not. The first is necessary, the second contingent.

The difference between the necessary and the contingent is recognized from the perspective of our theoretical framework. The decisive factor is theoretical rigidity, that is, being firmly tied into the conceptual web such that changing the one feature of nature has far-reaching consequences. A state of affairs is necessary to the extent that other aspects of nature depend on it. A statement is necessary to the extent that other components of our description of nature depend on it. There being no land animal faster than a cheetah is not at all necessary, since the discovery of a faster animal would not force any changes to theories of physiology, evolution, or anything else. Discovering a faster animal would be big news, but zoology books would only need to be lengthened a little bit, not significantly rewritten. The only substantive statement that would be negated is the contingent claim itself, the one that declares that there is no faster animal. But discovery of a causal signal faster than light would force fundamental revisions in mechanics, electrodynamics, atomic physics, cosmology and beyond. The claim about the speed of light is deeply and tightly embedded in the structure of our theoretical description of nature. This is how it is recognized as being necessary. A lot of other things couldn't be the way they are if casual signals didn't have this property. Causal signals have to be this way for nature to be the way it is.

The difference between a fact that is necessary and one that isn't is not intrinsic to the fact itself, and no isolated analysis will recognize necessity. Necessity is a global property of extensive connections among facts. Necessity is a kind of coherence. There is a real difference between facts like the speed of animals and facts like the upper limit of causal speeds. A lot depends on the latter, but little else depends on the former, and this difference of dependence is a real difference in nature. Science reveals this difference between the contingent and the necessary in terms of the coherence of theoretical connections and theoretical rigidity, and no rendition of scientific method that stops with testing hypotheses in theoretical isolation will be able to show how this is done.

Observation supplies the large assemblage of facts, but it does not reveal the interconnected structure associated with necessity. More observation gets us more knowledge, but not more understanding. We can know what is contingent in nature. We can also understand what is necessary, but only by attending to the non-empirical, derived links in our systems of theories.

Summary

So, what is science? And what is the basis of credibility of scientific results?

What makes science scientific? What work is being done by the word "scientific" in such weighty and authoritative phrases as "scientific evidence", "scientific testing", or, for that matter, "scientific method"? Consider the opposites. Unscientific evidence might be sloppy, incomplete, or not relevant to the ideas being tested. It is gathered and interpreted in ignorance of other evidence or theoretical context. To be scientific is to take full advantage of other relevant observations, relevant standards of observation, and relevant theoretical background. And it is the theoretical context itself that rules on what is relevant. The essences of science, in other words, is the explicit influence of this web of observations and theories.

Science is a process of assembling an interconnected structure of descriptive claims about nature. The structure and interconnectedness are key. Understanding scientific method requires understanding the links in a broad network of information, some of it from observation and some of it from theory. Science requires coherence among these claims. Lots of pieces must fit together, and no isolated analysis of one idea or one test will result in a good reason to believe a theory. Nor does a piecemeal analysis give you a good idea of how science works. Scientific method is global.

The observations and theories in a network of scientific information are linked and related in a variety of ways. Theories logically imply precise observations that are then checked in the process of testing. Theories explain observations that have already been done. Some theories explain other theories, securing inter-theoretical links in the network; this contributes to our understanding of nature. And some theories can be used as auxiliaries in the testing of others.

This is science, this coherent web of information. There must be numerous observations in the web, and it is important that more observations are regularly added. It's not just coherence in the web that makes each entry credible, it's coherence maintained as more and more observations are made.

Sometimes we find pieces of the web that no longer fit, perhaps because of new observations or because of rethinking some theoretical entries. In these cases, the scientific method is to reject, reinterpret, or reorganize entries in a way that maintains or enhances overall coherence. Sometimes whole inter-woven sections of the web must be rejected and replaced. But the decision and the process of doing this are not haphazard or whimsical. They are guided by objective standards of coherence. The small descriptive pieces are justified by their fit in the larger descriptive account of nature.

It's not uncommon to hear that there is no single scientific method. Not only is science given to indelible social influences that evade methodological constraints, but there is no one method or set of standards followed by all, or even most, scientists. But this is misleading. It's not that there is no scientific method; it's just not to be found in individual scientists. The method is a feature of a group of scientists, and it's the social interaction that holds it together.

Science is indeed a social phenomenon, and it is likely that no individual enacts the whole method. Individuals contribute parts, some hypothesizing, some testing, some doubting and revising, and so on. The whole method is an activity of the whole group.

Suggested Reading

Chalmers, A., *What is this thing called Science?*, 2nd edn, University of Queensland Press, 1982.
This is enjoyable and easy to read and it gives a pretty comprehensive survey of the things philosophers of science talk about.

Duhem, P., *The Aim and Structure of Physical Theory*, Atheneum, 1962 (original French edition, 1914).
Here is the classic source for the ideas of the interwoven nature of scientific knowledge, the necessity of auxiliary theories, and the indecisiveness of empirical testing.

Kitcher, P., *Science, Truth, and Democracy*, Oxford University Press, 2002.
Philip Kitcher is a highly respected philosopher of science. His account of science and objectivity, and the role of science in society, is clear and thought provoking.

Kuhn, T., *The Structure of Scientific Revolutions*, third edition, University of Chicago Press, 1996.
This is arguably one of the most influential books of our time. It introduces the ideas of a paradigm and paradigm shift, and raises the idea of non-rational influences in science.

Quine, W.V. and Ullian, J. S., *The Web of Belief*, Random House, 1978.
This is a short and very accessible book about the holistic network of our beliefs about the natural world.

Scheffler, I., *Science and Subjectivity*, 2nd edn, Hackett, 1982.
This level-headed discussion of subjective influences in science argues that objective results are nonetheless possible.

Toulmin, S., *Foresight and Understanding*, Harper & Row, 1961.
Here is wisdom and common sense about science and knowledge.

Weinberg, S., *Dreams of a Final Theory*, Vintage Books, 1992.
Steven Weinberg is a Nobel-Prize winning physicist with a literary skill for clarifying both the foundational ideas of physics and the structure of scientific method.